I0469736

Senior

Technology

Ministry

Mastering Mobile Devices

THE SENIOR TECH WAY

DENGUHLANGA
JULIA
KAPILANGO

MASTERING MOBILE DEVICES

Senior Technology Ministry

Mastering Mobile Devices

DENGUHLANGA JULIA KAPILANGO

DIVINE XTRAORDINARIAN

Detroit

600 E. Warren Avenue Detroit, MI 48201
sr.techs.ministry@gmail.com
Facebook: Senior Technology Ministry
www.srtechz.com

For GOD hath not given
us the spirit
of fear;
but of power,
and of love,
and of a sound mind.

Senior Technology Ministry

Our Senior Technology Ministry is committed to patiently train senior citizens at their level of technological learning. We train you or your parents/grandparents/great grandparents in using mobile devices. We believe with this form of training it leads to improving their quality of life, enhancing their social and civic engagement both at beginner, intermediate and advance learning level.

We are committed to decreasing new technology intimidación and social media avoidance. Through this approach we will increase many older adult end users to shop, chat, share and bank online using their station technology, mobile devices and the worldwide web.

Contents

Forward

by
Shirley Northcross
Retired Physical Education Teacher, Detroit Public Schools

Our oldest son who works at AT&T decided that it was time that we upgraded from our wonderful, easy to use Flip Phone and transfer over to the challenging Apple iPhone! We were pleased that he cared but as we listened to the directions to do this and to do that, everything was over our heads!

Eager not to be left behind in the ever changing Techie world. I approached Denguhlanga at church. She gave me a "BRIEFING" that I understood! WOW FACTOR SAT IN and I was eager for more!!!

My friend and fellow church member was standing there at the time. She had a similar experience with her family. Both of us were soaking up information faster than Denguhlanga could transmit it to us.

If we asked her to repeat it, she never responded impatiently, "I told you that once." She simply said, "let's try explaining it another way ---WELL! This turned into a marriage made in heaven and the Senior Technology Ministry Class was born!! I soon began sharing our experience and recruiting all my Senior Technology challenged buddies and we were off and running with Senior Tech Class 101.

This technology instruction is important as Seniors do not want to be left behind. We want to be aware of those developments that will improve the quality of our lives and that prepares us for the future.

Denguhlanga provides and incredible service ministry because she is very talented to be able to meet each of us where we are and has the patience to get us to the finish line.

She knows her technology forward and backwards. She brings GOD into the picture with joy.

Learning about the mobile devices is worth the effort because family and friends will appreciate, respect and love our efforts. We are staying in touch with the 21st century and enhancing our ability to function in today's world

Special Thank You!

Special Thank you to my Mother Julia A. Kapilango, my Father Adao Jose Domingos Kapilango, Antoine D. Maddox Bey, Jacqueline Kapilango, Brandice Atoi Neely, Binga Smith, Dr. Rev. Nicholas Hood Sr, Rev. Dr. Nicholas Hood III, Delores Hilson, Dr. Marilyn French Hubbard, Dr. Kevin M. Turman, Malik Yakini David Rambeau, Barbra Williams, Charity E. Hall, Will Amos, Shirley Patton, Tim Smith, Tobius Smith, Shirley Northcross, David Northcross, Joyce Wortham, Gloria Lowe, The Clark Family, Tamiko Leonard, Kelley Anne Jones, Henrietta Carson, LaTasha Washington, Dr. Rev, Georgia A. Hill, Donavan, Shari Sertima, Natalie Sertima, Gloria Jean Cooley, Judith Harvey, Claire Neal, Dr. Brenda McGadney, Dr. Shirley McRae,

Dr. Yvonne Catchings, Patricia L. Millender, Dr. Cora Collins,
Dr. James Collins, Michael Daniels, Dr. Marsha Malone Thompson, Mamie Cokley, Teresa Barmore, Robert Polk, Janet Whitaker, David Whitaker, Brenda Peek, Errol Griffen, Joan Galloway Blount,
Dr. Tommie Johnson, Mary Fritz, Debra Copeland, Lois Murphy,
Dr. Ronald Williams, Wayne State University Institute of Gerontology Outreach Department, Detroit Area Agency on Aging, Barbara Leslie, Dr. Ann Smith, Dr. Robbya Green – Weir, Destine Reese Collaborative, Inc.

I sincerely thank you all!!!!

"There is a fountain of youth: it is your mind, your talents, the creativity you bring to your life and the lives of people you love. When you learn to tap this source, you will truly have defeated age."
— **Sophia Loren**

I

In The Beginning

Tons of people ages 54 and up are turning in their old cellphones, bulky CPU's and monitors.

Why? There are many reasons why this massive movement of senior techs is taking place.

Seniors are hipper and much cooler than those who came before. Seniors are taking control of how they will grow old. Many seniors are continuing to challenge their brains - way after retirement with mobile devices.

Now a days, seniors have smartphones, androids, tablets, laptops and MacBook's. All of these devices are transportable an easy to manage.

Managing your mobile devices at first may seem a bit intimidating or confusing. You pick which one works for you.

However, when you first get your mobile device, you are pretty much left by yourself. You are left by yourself to figure out this fancy – high tech gadget.

You must figure it all out. You must figure out how to open up your phone.

What a sim card is. Where your battery is located. How to turn your phone on and off.

Where to plug your charger to charge your mobile device.

It is so much you must learn before you even before you even get started on making a simple phone call.

Why? Is it because we do not want to look stupid. Or, could it be we do not want to appear old and dumb. Wait, here is a good one. I do not need these fancy phones. I will use my handy 1999 style flip phone until I die.

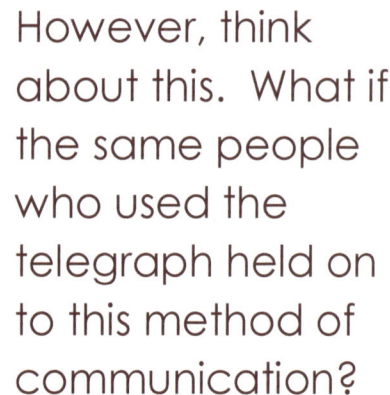

However, think about this. What if the same people who used the telegraph held on to this method of communication?

Where would technology be? I will tell you. We would be tapping Morse code to one another right now. Morse code an alphabet or code in which letters are represented by combinations of long and short signals of light and sound. (Google, 2015)

Anyhow, this book is written especially with you in mind. You want to understand how to read your current data plan. Maybe you want to get a bit more social via Facebook.

Here's a good one. All your children have moved away and taken your grandchildren with them. How will you see them?

How will you talk to them?

Skype or FaceTime! Whatever level of learning you are on as a Senior Tech, Senior Technology: Mastering Your Mobile Devices will provide some very helpful information to encourage you.

We all want encouragement. So take the first step. Read each page with calmness, confidence, and common sense.

With these three powerful elements you will master all your mobile devices as a Senior Tech.

"Keeping up the appearance of having all your marbles is hard work, but important."
— Sara Gruen

II

Let's Get Started!

Many of us go out and by the first Mobile device we see. Why? Because the sales rep suggested this mobile devices for our current lifestyle.

Our children purchase the mobile device for us without asking us. The senior citizen organization is partner with the mobile device provider.

Whatever it is, go to your local library – read magazines or newspaper articles – ask your senior technology instructor or someone you really trust to help you pick out a new mobile device.

All mobile devices are not the same. Some mobile devices are easier to use than others. There are mobile devices that have apps. Mobile devices that are made specifically for seniors.

Picking out your next mobile device should have the same consideration you take when purchasing a new car. Both are mobile. Here is a list of mobile devices every senior should own or have access to.

Laptop/Notebook – A computer that is portable and suitable for using while traveling

Wireless Printer – A printer without a cord connection that works with a proper WI-FI

Smartphone – A cellular phone that performs many of the functions of a computer, typically have a touch screen interface (you can touch the screen with your figure), internet access, and operating system capable of running downloaded applications.

Android phone – An open – source operating system used for smartphones and tablet computers

Tablet/iPad – A handheld tablet computing device with touch screen interface that accepts input directly onto an LCD screen rather than via a keyboard or mouse

Now that we have some understanding about mobile devices, let's move unto acquiring a proper data plan.

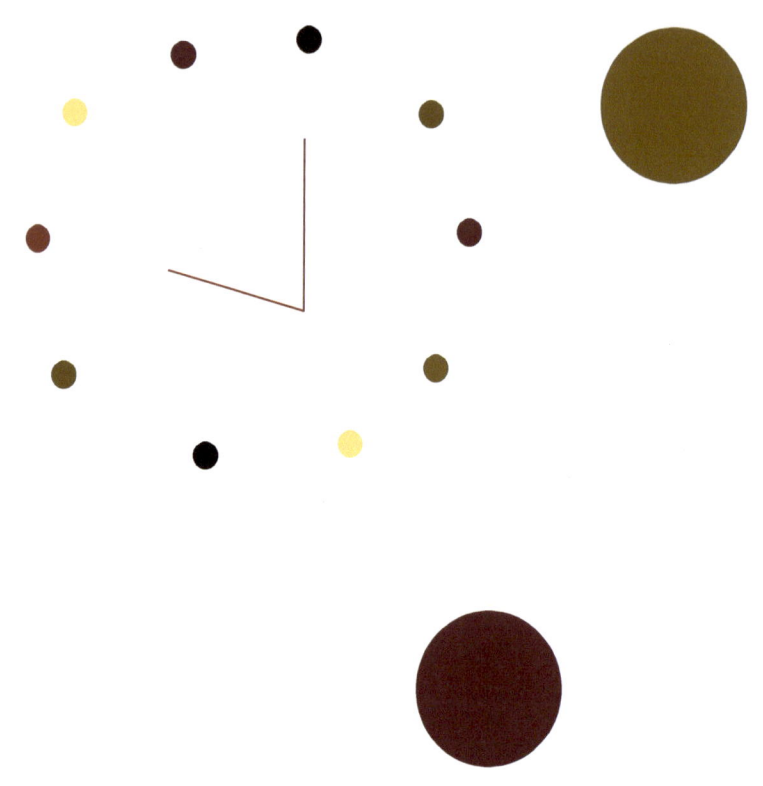

"Age has no reality except in the physical world. The essence of a human being is resistant to the passage of time. Our inner lives are eternal, which is to say that our spirits remain as youthful and vigorous as when we were in full bloom. Think of love as a state of grace, not the means to anything, but the alpha and omega. An end in itself."
— Gabriel García Márquez,

III

PICKING THE RIGHT PLAN

Selecting the right data plan for you depends on a few things. The first thing is, how many people and what types of mobile devices will use the one plan.

Knowing how many people and mobile devices using the data plan helps you understand how much data you will want to purchase every month.

According to CTIA data, consumers with smartphones —such as an iPhone, Android, or Blackberry—use an average of about 800 MB of data per month. That's far less than the standard 2 GB of data recommended by cellphone companies. And 800 MB is no small amount. In fact, a quick analysis shows it's more than

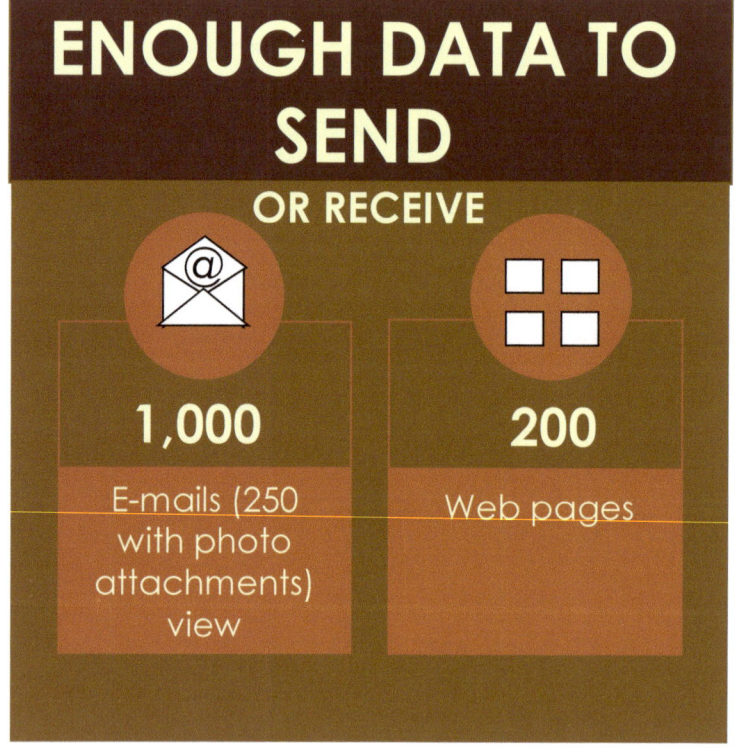

ENOUGH DATA TO SEND

OR RECEIVE

1,000

E-mails (250 with photo attachments) view

200

Web pages

50

photos on social media, download

10

apps, songs or games, stream

10

hours of music, and watch

60

minutes of YouTube videos

The data plan can be through a contract or a no contract provider.

No matter what carrier (Ex. AT&T, Boost Mobile, Verizon, MetroPcs, and Consumer Cellular) you're with, be sure to check with it frequently, at least during the first few months of your contract, to monitor your data usage.

You may be able to access your data usage info by texting your provider, logging into your online account, or downloading a cellphone app. (Citizens Utility Board, 2015)

Listed below are a few definitions to help you on your senior technology journey:

Data is distinct pieces of information, usually formatted in a special way. All software is divided into two general categories: data and programs. Programs are collections of instructions for manipulating data. Data can exist in a variety of forms -- as numbers or text on pieces of paper, as bits and bytes stored in electronic memory, or as facts stored in a person's mind.

Data Plan is a monthly subscription to a cellular or other wireless carrier for the transfer of data over its network. It may be an unlimited-use plan or based on the actual amount of data transferred. It is a monthly subscription to a service that delivers database content, real-time data, news or other information.

Usage is the act, manner, or amount of using data on mobile devices

Giga Bite

A gigabyte (GB) is a measure of computer data storage capacity that is roughly equivalent to 1 billion bytes. A gigabyte is two to the 30th power or 1,073,741,824 in decimal notation. The term is pronounced with two hard Gs. The prefix giga comes from a Greek word meaning giant.

Mega Bite

The megabyte is a multiple of the unit byte for digital information. Its recommended unit symbol is MB, but sometimes MByte is used.

Through learning your different needs for a data plan, you will be able to save time spent on the telephone with your service provider. You will also lower your stress level by understanding how to turn mobile devices on your plan off.

HELPFUL HINT:

Whenever you are at home or in a location that has free WIFI make sure to connect your mobile devices to that connection.

25
55
35
45
65
75

Nobody grows old merely by living a number of years. We grow old by deserting our ideals. Years may wrinkle the skin, but to give up enthusiasm wrinkles the soul. ~Samuel Ullman

IV

GET YOUR SOCIAL NETWORKING ON

In this day and age social networking is the big thing. If you don't have a Facebook Page or a Twitter Handle you are missing out on all the fun happening on social media.

Many of your **greatest** heart throbs from your childhood are very socially active (SA) on the World Wide Web. I know, there are lots of scary things happen with your identify possibly being stolen. No worries, there are several different types of security measures that you can take to get SA.

You can post a picture to your profile of a favorite hobby, animal, plant or even a record cover. You do not have to post a picture of yourself on your profile page. Also, create a junk mail email address account. You can use this email account to set up your social media account (SMA).

Make sure you get **creative** coming up with your username such as Classy Claire or Sweet Strawberry.

As you get comfortable with being SA you will begin to use #Hashtags to place emphasis on some of your words. #Yougetit.

You will also have access to direct messaging. This allows you to send people you have connected with via SM direct messages to their personal pages.

Listed below are some useful terms and steps to get you socially active on your senior tech journey via social networking.

Facebook is a popular free social networking website that allows registered users to create profiles, upload photos and video, send messages and keep in touch with friends, family and colleagues.

Use:

1. Go to www.facebook.com.

2. If you see the signup form, fill out your name, email address or phone number, password, birthday and gender. If you don't see the form, click Sign Up, then fill out the form.

3. Click Sign Up.

4. After you join Facebook, there are a few ways to find your friends or invite them to join: Search for friends, Import your contacts, and Invite friends individually

5. You can share stories from the top of your Timeline or your News Feed.

6. Your profile is your collection of the photos, stories and experiences that tell your story. Your profile also includes your Timeline.

7. Use your Settings to manage basic account preferences. You can edit your name or email info, change your notifications preferences, turn on extra security features and more.

Direct Messaging

A direct message (DM) is a messaging function in Twitter that allows a user to send a private message to a specific user. Unlike the normal tweets that can be seen by all the user's followers, a direct message can only be read by its recipient.

Instagram

Instagram is a free online program and social network that enables users to take, edit and share photos with other users via Instagram's own platform, email, and social media sites including Twitter, Facebook, Tumblr, Foursquare and Flickr.

Use:
1. Take a photo/video and choose a filter.

2. On the screen you see after choosing a filter, type your hashtag in the Caption field (ex: #IAMSrTech).

3. If you want to tag a post you've already uploaded, edit the caption or include your #hashtag in a comment on your photo.

Hashtag
Hashtag is a word or phrase preceded by a hash sign (#), used on social media sites such as Twitter to identify messages on a specific topic.

Learning how to navigate using social media is fun and very engaging. However, one of the most important things to **remember** is, if you are going to post a picture of yourself. Make sure the selfie is adequate.

S
E
FAITH
L   H
CONFIDENCE
      P
      E

You are as young as your faith, as old as your doubt; as young as your self-confidence, as old as your fear; as young as your hope, as old as your despair. ~Douglas MacArthur

PUTTING YOUR BEST SELFIE FORWARD

A selfie is a photograph that one has taken of oneself, typically with a smartphone or webcam and shared via SM.

Lots of times bad selfie's are posted. Selfie's that look like police mugshots - **BAD**.

To take a correct selfie, one must flip camera on phone or stand in the mirror looking at self.

Consider taking only a shoulder up headshot. This is if you are conservative.

However, you can also take full body shots. Hey no butts to the camera's. LOL. Remember to smile and get your good side. Selfies are not the only way to get your glam on our share images of yourself. There are other apps such as FaceTime and Skype that **empower** you to have live face to face conversations (F2FC).

Listed below are a few steps to help you with F2FC:

FaceTime

A video telephone / video chat service somewhat similar to Skype and Google Hangouts that makes it possible to conduct one-on-one video calls between newer Apple iPhone, iPad, iPod touch and Mac notebooks and desktops.

Apple's FaceTime service is free to use but does require an Apple ID and a Wi-Fi connection, although future versions of FaceTime may also work over 3G and/or 4G connections, and several apps are currently available that make FaceTime over a 3G connection possible on a jailbroken iPhone.

USE:

1. Launch the Phone app from the Home screen of your iPhone.

2. Tap the Contacts tab at the bottom

3. Tap on the contact you'd like to FaceTime.

4.   Tap on
FaceTime near
the bottom of
their contact
card.

Skype

Skype is a computer program that can be used to make free voice calls over the

Internet to anyone else who is also using Skype. It's free and considered easy to download and use, and works with most computers.

Once you download, register and install the software, you'll need to plug in a headset, speakers or USB phone to start using Skype.

Step 1:⟶

Install Skype. Skype is a
free app for both Android
and iOS devices.

Step 2: ⟶

Set up Skype. Android:
Once Skype for Android is
installed, tap on the app
to open it.

Step 3:⟶

Make a Call. You can
make two types of calls
using Skype: Skype-to-
Skype calls and Skype-
to-Phone calls.

So if the grandkids are in Ghana or your Boo is in rehab, you can still talk and see each other in real time.

Real time usage with your mobile devices is limitless. There is always something new happening with technology. I **encourage** you to continue to grow with technology.

Few Extra Tips

#1
#2
#3
#4

Let me give you a few extra tips.

Tip #1 - Update your current mobile devices every 2-4 years depending on where technology is going.

Tip #2 - Never get frustrated with your mobile devices.

Tip #3 - Please do not punch your phone screens.

Tip #4 - Make sure you purchase a protective case and screen covering for your mobile devices (i.e iPhone, Smartphones, ect).

Tip #5 - Always have a charger for your mobile devices. There are all kinds of chargers available. There are car chargers, solar chargers and plug-ins. There is no excuse to run out of juice.

Tip #6 - Always make sure your phone is turn on before you say it is not working.

Last but not least,

Tip #7 - Remember to keep **calm**, **be confident** and use your **common sense**.

Enjoy your mobile devices **Senior Techs.**

GLOSSARY
A - B - C - D

App - is an abbreviated form of the word "application." An application is a software program.

App Store – is an online store to sell, purchase and download apps for iPhones, iPads, iPod touch and Mac OS X PC

Blog – a regularly updated website or webpage, typically one run by an individual or small group that is written in an informal or conversational style

Cloud – provides users and enterprises with various capabilities to store and process their data in third party data centers. (Think of a file cabinet that stores limitless information)

Data – distinct pieces of information, usually formatted in a special way. Data can exist in a variety of forms – as numbers or text on pieces of paper, as bits and bytes stored in electronic memory, or as facts stored in a person's mind.

GLOSSARY
D - E - F

Data Plan – monthly subscription or prepaid card to a cellular/wireless carrier that delivers database content, real-time data, news or other information

Direct Messaging – a function that allows users to send private messages to a specific user with in a specific social media site

Download – copy data from one computer system to another, typically over the internet

Email - is a system for sending messages from one individual to another via telecommunications links between computers or terminals using dedicated software

Facebook – a free social networking website that allows registered users to create profiles pages, upload photos and videos, send messages and keep in touch with family, friends and colleagues.

GLOSSARY
F - G - H

FaceTime – a video telephone service that makes it possible to conduct one – on – one video calls

Giga Bite – is a measure of computer data storage capacity that is roughly equivalent to 1 billion bytes

Google Play – online store for purchasing and downloading apps, music, books, movies and similar content for use on android – powered smartphones, tablets, etc.

Hashtag – a word or phrase preceded by a hash sign (#), used on social media sites such as Twitter and Facebook to identify messages on a specific topic

GLOSSARY

I - L - M

Instagram – a free social network that enables users to take, edit and share photos with other users

Login - is an act of logging in to a computer, database, or system.

Logout - means end a session at the computer. For personal computers, you can log out simply by exiting applications and turning the machine off. On larger computers and networks, where you share computer resources with other users, there is generally an operating system command that lets you log off.

Mega Bite – is a multiple of the unit byte for digital information

GLOSSARY

P - S

Pinterest – a social network for sharing and collecting categorized pictures uploaded or from online

Profile Page – information about an individual user within the social media app, typically includes their name, education background, marital status, ect

Selfie - is a photograph that one has taken of oneself, typically one taken with a smartphone or webcam and shared via social media.

Skype – is a computer program that can be used to make free voice calls over the internet to anyone else who is using Skype on their laptop, tablet, smartphone, android phone or desktop computer

GLOSSARY

T

Text Messaging - or texting, is the act of composing and sending brief, electronic messages between two or more mobile phones, or fixed or portable devices over a phone network. The term originally referred to messages sent using the Short Message Service (SMS)

Twitter – a free social networking microblogging service that allows registered members to broadcast short posts called tweets

Tumbler - a free social networking microblogging service that allows users to post multimedia and other content to a short-form blog

GLOSSARY

U

USB - short for Universal Serial Bus, an external bus standard that supports data transfer rates of 12 Mbps. A single USB port can be used to connect up to 127 peripheral devices, such as mice, modems, and keyboards. USB also supports Plug-and-Play installation and hot plugging.

Upload – transfer data from one computer to another, typically to one that is larger or remote from the user or functioning as a server

Ustream – is a website that allows members to broadcast live streaming video and archiving on the internet to registered users (Watch our Sunday Live Service at http://www.ustream.yv/channel/ply mouthdetroit)

GLOSSARY

W - Y

Web Browser – is a software application for retrieving and presenting information resources in the World Wide Web such as Safari, Internet Explorer, Google Chrome, and Firefox

WordPress – is a free, open source publishing software that can be installed locally on a web server and viewed on a proprietary web site or hosted in the cloud and viewed on the WordPress website

YouTube – is a video sharing website that enables viewers to watch, save and upload entertainment of all genres

References:

Citizen Utility Board. (2015). Telecom Data Guide. Retrieved from www.citizensutilityboard.org/ciTelecom_DataGuide.html#

Dictionary Reference. (2015) Definition of email. Retrieved from http://dictionary.reference.com/browse/email

Getenger. (2015). What is a text message or sms? Retrieved from http://resources.getenger.com/r/what-is-a-text-message-or-sms-message/

Oxford Dictionary. (2015). Definition of login. Retrieved from http://www.oxforddictionaries.com/definition/english/login

Oxford Dictionary. (2015). Definition of selfie. Retrieved from http://www.oxforddictionaries.com/definition/english/selfie

Search Mobile Computing. (2015). Definition of App. Retrieved from http://searchmobilecomputing.techtarget.com/definition/app

Time. (2012). 92 Teen Text Terms Decoded for Confused Parents. Retrieved from http://www.techland.time.com/2012/05/03/92-teen-terms-decoded-for-confused-parents/

Webopedia. (2015). Definition of Logout. Retrieved from http://www.webopedia.com/TERM/L/log_out.html

Webopedia. (2015) Definition of USB. Retrieved from http://www.webopedia.com/TERM/U/USB.html

143 – I love you
4EAE – for ever and ever
4YEO – for your eyes only
AND – any day now
AFAIK – as far as I know
ATM – at the moment
BOL – be on later
FYS – feed your soul
FWB – friends with benefits
GIGR8 – God is great
GR8 - great
HF – have fun
IDK – I don't know
IKR – I know right
IRL – in real life
K - ok
LOL – laugh out loud
Lmbo – laugh my butt off
MM – music Monday
NAGI – not a good idea
P4U – praying for you
PG – praises God
ROTFL – rolling on the floor laughing
SMH – shaking my head
TIA – thanks in advance
TMB – tweet me back
TTYL – talk to you later
TYT – take your time

TEXT

TALK

_____

ABOUT
The
Author

Denguhlanga Julia Kapilango has been working with technology since the age of nine. She learned from technology masters such as Dr. Ronald Williams, former superintendent of Highland Park School District and Ms. Lois Murphy, former computer science department head for Detroit Public Schools.

Denguhlanga Julia is a web advocate. E-Marketing specialist, twitter practitioner, social media scholar, digital design aficionado, P2P connecting guru and a passionate transformational entrepreneur. She holds degrees and certifications from Baker College, Baker Center for Graduate Studies, YouTube Creator Academy and Google Academy.

Denguhlanga Julia is the founder and instructor of the Senior Technology Ministry also known as STM. Classes are held weekly at Plymouth United Church of Christ. She founded STM on her passion to serve older adults.

Denguhlanga enjoys sharing her skills and training methods to senior techs. She provides a nurturing and confident building environment for senior techs to take control over their different mobile devices. STM is for older adults ages 62 and up who need a personalized simplified training on using laptops, iPhones, Androids, iPads and Flip Phones.

Denguhlanga's mission with STM is training senior techs to learn and keep connected through understanding the history of technology, basic coding language/ terminologies, introductions to new apps for teaching, setting up LIVE Streaming Events, group networking for meetings/family programming, online banking/check

SENIOR TECHNOLOGY MINISTRY
A guide for adults 60 and older

Now that you have read this spiritual technology ministry by Denguhlanga Julia Kapilango ----What is your next step?

Our Senior Technology Ministry solves the numerous statistics affiliated with older adults in the areas of isolation and confinement, mental and physical exercise, socialization, digital divide, and digital illiteracy.

We offer several 4 to 8 weeks Senior Tech Ministry trainings' that equip older adults with high levels of confidence, independence and access to answer their own technology questions, curiosities and concerns. Instruction is limited to 8 to 10 students per session. Our Senior Tech Ministry training format allows older adults a comfortable and stress free one-on-one interaction.

For more information about the Senior Technology Ministry, or to find the class nearest to you, please visit: www.srtechz.com.

Senior Technology Ministry®
Mastering your mobile devices
Senior Technology Ministry / Divine Xtraordinarians Publishing
600 E. Warren Avenue
Detroit, MI 48201
Phone: (313) 434-8350

www.ingramcontent.com/pod-product-compliance
Lightning Source LLC
Chambersburg PA
CBHW040843180526
45159CB00001B/292